You'l Ree...

riddles by Billy Aronson and Ruby Mae

illustrated by Bill Mayer

HARCOURT BRACE & COMPANY

Orlando   Atlanta   Austin   Boston   San Francisco   Chicago   Dallas   New York
Toronto   London

Where did the boy
take his pet pig?

3

To a <u>ham</u>usement park!

4

When a pony has
a cold, what does
it sound like?

A <u>Hoarse!</u>

What will your dog say
is the most important
part of the house?

What did momma pig put on baby pig's boo-boo?

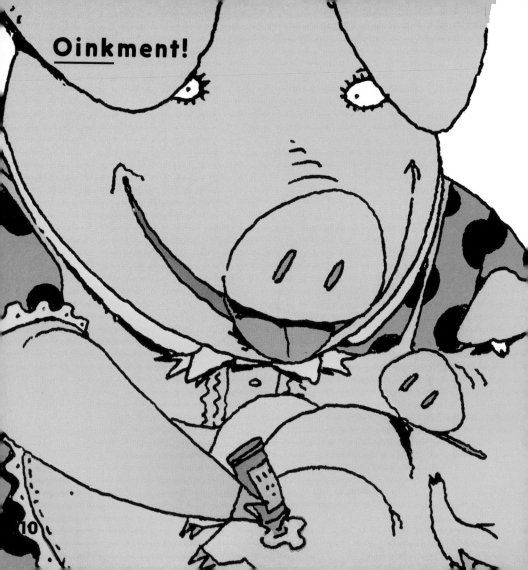

What noise does a mouse's favorite toy make?